T0392288

Earth's Resources

By Adam Elfadl

To order additional copies of this book, contact:
Xlibris
844-714-8691
www.Xlibris.com
Orders@Xlibris.com

ISBN: 978-1-6698-6594-0 (sc)
ISBN: 978-1-6698-6593-3 (hc)
ISBN: 978-1-6698-6592-6 (e)

Library of Congress Control Number: 2023902129

Print information available on the last page

Rev. date: 02/07/2023

Earth's Resources

by Adam Elfadl

MEET THE YOUNG WRITER

My name is Adam Elfadl, I am 9 years old,
I was born on January 12,2013.

Reading books was one of my favorite things to do at early age. Besides reading and writing, I started playing piano when I was 4 years old at Forté Music School Toledo OH under the supervision of Mrs. Melanie Zientek, to continue learning Recorder and Basketball.

I was inspired to make this book when I asked my mom what our kitchen table was made of, and she answered granite!
At that time, I was very curious, what is granite? and how a big rock could be transformed to a big shiny countertop!
I started researching and found myself ready to share my findings.

The biggest reason behind publishing this book is to inspire kids that any dream can be reached by working towards it, and to remember that success is everyone's innate desire.

MEET SOME OF OUR WONDERFUL EARTH RESOURCES

Gems & Minerals – Rocks – Fossils – Plants

There are a lot of gems present in the Earth and in space, most are very common, while others are increasingly rare.

The majority of what we find are minerals, which make up rocks, therefore rocks can only be broken down into minerals and nothing smaller than a mineral can be in a rock.

Gems are not considered ores such as copper or iron, they are minerals that are often made from volcanoes and can also be found underground, formed under extreme pressure.

Gems are often formed from minerals that get stuck in a forming rock. This happens when pressure pushes a rock into another at such high pressures that they form into one. The gem must be inside the rock or where the inside will be when the rock is done forming. This only happens to igneous and sedimentary rocks.

Minerals

and ores are incredibly useful and are used in all sorts of ways such as the manufacturing of computers, vehicles, buildings, and even when you see your mom's jewelry, you'll probably see gold and silver.

Copper

has been used for thousands of years to make both items needed for basic survival and fashionable jewelry, often designed in a way that honors respected ancestors or believed gods.

For millennia, people used ax heads made from copper to hunt and fight, they also used spearheads which they could shoot from arrows or attach to spears.

They made shields from copper and copper oxide and sword hilts from wavy unstable copper, they also used bronze copper for masks to commemorate their fallen rulers.

MALACHITE

Malachite is a striking green mineral that is used mostly for fashion, jewelry, healing, and painting.

Malachite was used impressively in the construction of The Mayan Mask, which was the funeral mask made for the red queen from Mexico, and part of the building of the grand Kremlin palace in Moscow.

GRAPHITE

Graphite is a useful mineral for pencil lead. It is also used in the investigation of crime scenes; when the police check for fingerprints, they brush the contaminated item with powdered Graphite to reveal the mark.

Graphite is also a good conductor of heat and electricity. Sometimes, when power plants lose uranium, they either turn to graphite or unstable silver or toxic silver oxide.

Graphite can be made into carbon fibers, which can be used to make stunt helmets and other high-performance parts in the motor industry. In the future, graphite will be used for flexible touch-screen glass.

Graphite epoxy can be used for aircraft fuselages.

Platinum

is mostly used for fashion and currency; platinum can be used for rings and bracelets but can be quite expensive.

The cat masks from Ecuador were made from gold with platinum eyes. Platinum can be a valuable item in currency, 10 tons of platinum can be worth 1 ounce of gold bullion.

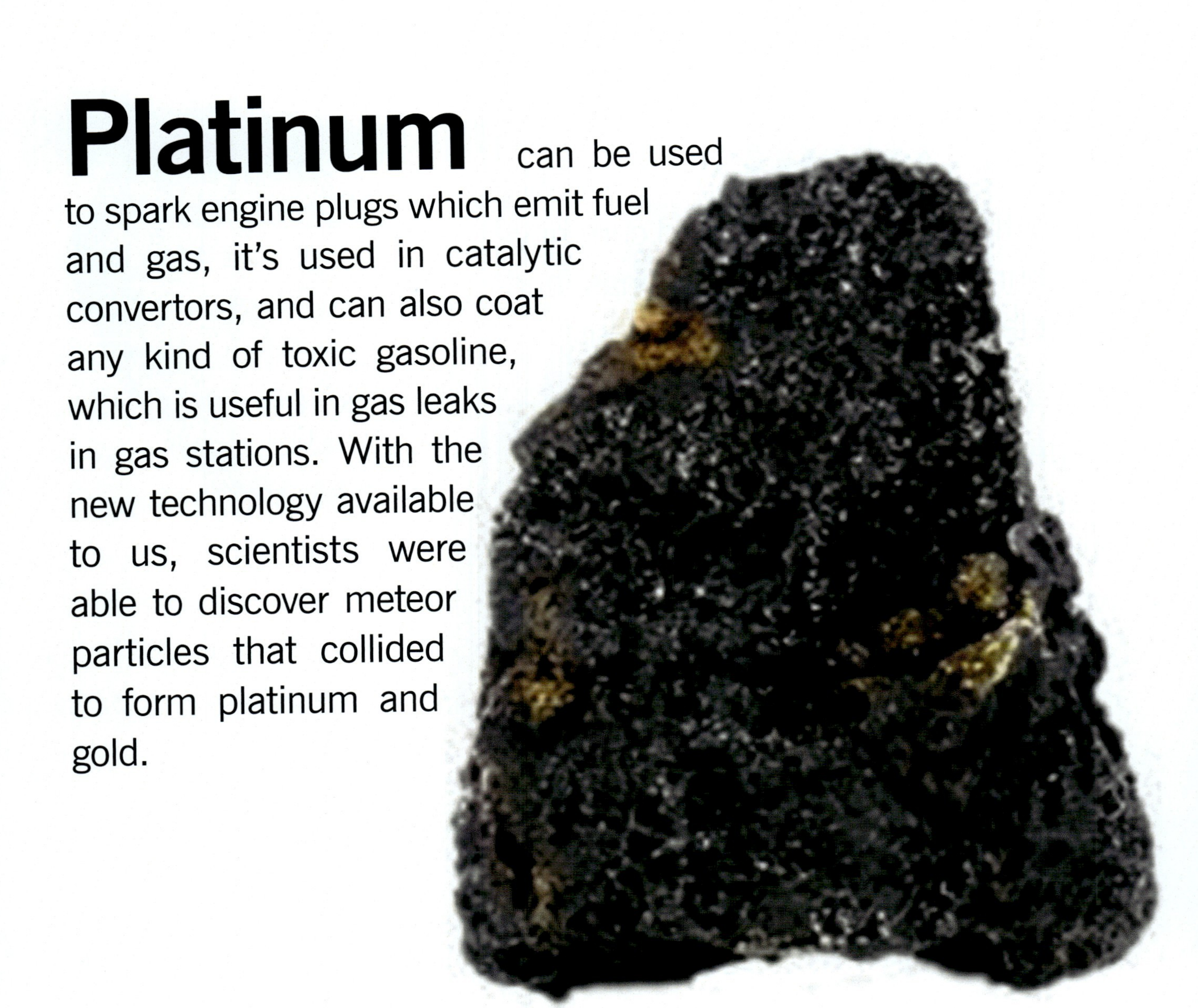

Platinum can be used to spark engine plugs which emit fuel and gas, it's used in catalytic convertors, and can also coat any kind of toxic gasoline, which is useful in gas leaks in gas stations. With the new technology available to us, scientists were able to discover meteor particles that collided to form platinum and gold.

Iron

Is a very common metal found on many different planets. It can be treated in many ways and has a multitude of uses. Cast iron can be made into pots and pans that, when put in extremely hot conditions, will still maintain their form and structure. Iron can be melted and then hardened to make hard iron which can make bridges, and tall structures.

Wrought **iron** is more pliable and was often used for building furniture, the Eiffel Tower was made from wrought iron. Steel was eventually discovered when Small amounts of iron were melted and compressed in a certain form to make steel which can be used to make impressive, gravity-defying structures, for Example, most of **Sydney** opera house is made of steel.

Gold

is a valuable item in currency. Gold has such an attraction that it can cause hundreds of thousands of people together in a desert to dig for it.

This was what happened in America in the 19th century, some were lucky and had tools and labor, while others had to use a pan and dig the sand in the search of gold.

Silver

is a useful item, not only in currency, but in many other ways in modern life. Silver was used to trade in the early years when gold was not used or found.

Silver was rare, so it took its place as a currency, and many people still use silver bars to preserve their wealth.

Silver is a good conductor of electricity, so it can be used in circuit boards. Silver can be made into silverware to make spoons, forks, and other kitchen appliances.

Silver was used for early photos instead of film and glass slides.

Olivine

is the most common mineral found in the earth's upper mantle. Olivine can also be found in high-temperature metamorphic rocks, lunar basalts, and some meteorites.

Olivine can be made into cut gemstone either by man-made tools or weathering, although weathering \was used in many cultures such as in Egypt, where many people made symbols with stable, weathered olivine to make words in the pyramids.

Olivine

is mostly found underground or in toxic lakes, but it can be found in a metamorphic stage when olivine is in a pool of lava in a volcano. When this is cooled, it will make an olivine 'bomb' which can explode.

Olivine w a s f o u n d in a meteor site which previously had no trace, which means particles in space consist of the material.

Zircon

is a mineral that can be found in many colors and can be used for many reasons. It is an ancient mineral, formed when decomposition emits energy which turns to radiation, but then gets solidified in space by the lack of heat. Zircon has a high melting temperature of about 2190°C (3974°F) and remains stable until 1750 °C (3182°F).

Zircon is made of little minerals called zirconium which can radiate and can be dangerous in their unstable form, but that only happens during its formation.

In the past, zircon was used for fashion and to this day can be found only in rare occurrences as just a little zircon is used in today's modern jewelry.

Many people confuse zircons with diamonds, but I personally think if you find it faster it's probably fake :)

Zircon can be made into many plastic kitchen appliances which can be very useful.

Chromite

doesn't just sound like a computer store, it can be useful for building them, as many buildings need the mineral for their supporters. It can also be used to make a metallic looking metal which can be put on bikes and other transportation.

Chromite was used for yellow paints and shaded or unshaded yellow chromite dye.

Magnetite

is a useful mineral because it is magnetic. Many people bring compasses on their trips because if they get lost, they can navigate with the compass. The white and red parts are made of magnetite and react to the magnetic pull of the earth's poles. The red shows the way you are supposed to go to reach the north, and the white hand points to where the south is.

Many paperclips are made with magnetite so when a magnet is near them, they will attach to it. Other modern-day uses include abrasives and pigments.

Chalcedony

is a mineral that was used in the past for survival and natural generation. Chalcedony can be found in fossil form, when something is compressed into another, it turns into a fossil. Also, it was used as a traditional knife in Mexico, people used the knife to hunt or cook.

It is a common item used in jewelry as it is said to have healing properties.

BORAX

is a mineral that is commonly used all over the world. When borax is crushed, the powder can make a dye that can be put on trees in the winter to stop them from dying. Borax was widely used for different cleaning products. It was fast, but wasn't the best at cleaning the big Messes, so after a year or so it was used for other things.

It was used in many laundry detergents until it was found that it could be harmful to your health. When hardened, borax becomes bulletproof, which was good and bad, because armies could use the vest but would not tell anyone, it saved many lives, but also ended many.

Many people tried to start businesses supplying borax. A popular one was called the 20-mule team, which was donkeys transporting Borax to companies.

Bauxite

is a mineral that has been used for many inventions. Fireworks use Bauxite, and when making the famous sputnik 1, many people weren't sure what to use, so they made a rare steel using bauxite, which was the most expensive product to make at the time. It pretty much bankrupted the whole country, which kind of led to the leader's depression after seeing the longest receipt he had ever seen.

Fluorite

is a beautiful mineral, so it was often used for jewelry and decoration, which bought people's attention to first class jewelry. The most amazing thing about fluorite is that when it is exposed to light waves in the dark, it will start glowing. As I said, Fluorite can be used for decoration, it started when people started to buy so much fluorite jewelry, and the shops started getting greedy and wanted more, so they called the mining companies and kept asking for more Fluorite.

But in the mines, the workers were becoming tired and ill. In this condition, they kept leaving to search for other jobs, so in the end, the mining companies shut down.

The shops didn't know what decorations to make with the fluorite that they had, so they crushed the fluorite, melted it, and put it in a cool place. The final look was so colorful that they put it in glass products and made many more.

Barite

is often found in hydrothermal veins and as veins in limestone. In the past, people used barite to make stones that could be burned and were said to bring magical creatures into the room to possess the rich and greedy. People used their tales to warn people, mostly children, that if they were lying, the creature would get them.

Barite can also be used for x - rays. When a human swallows liquid barite, the liquid will cover the organs so the doctors can better see any affected areas.

ROCKS

From sedimentary to metamorphic, rocks are amazing and very interesting. They are also super fun to find and have many different textures and uses. They were often used in the past to show off status and wealth, like fine marble furniture, sculptures, and buildings. Rocks are useful, and you might not know this, but the whole Earth is a cluster of huge rocks.

This happened in the making of the universe, clusters of rocks would be pulled to different locations of concentrations, and each one would have different amounts of gravity, which meant some would act as a magnet and pull others closer to form earth-sized clusters.

Hardness & Luster

The **Mohs** hardness scale is a way of classifying rocks although it only consists of 10 minerals the advanced will hold almost 89% of the known mineral's, talc is the weakest and diamond is the Hardest, when a mineral a lower level in the **Mohs** scale is scraped with a higher one will scrape the weakest one on the scale.

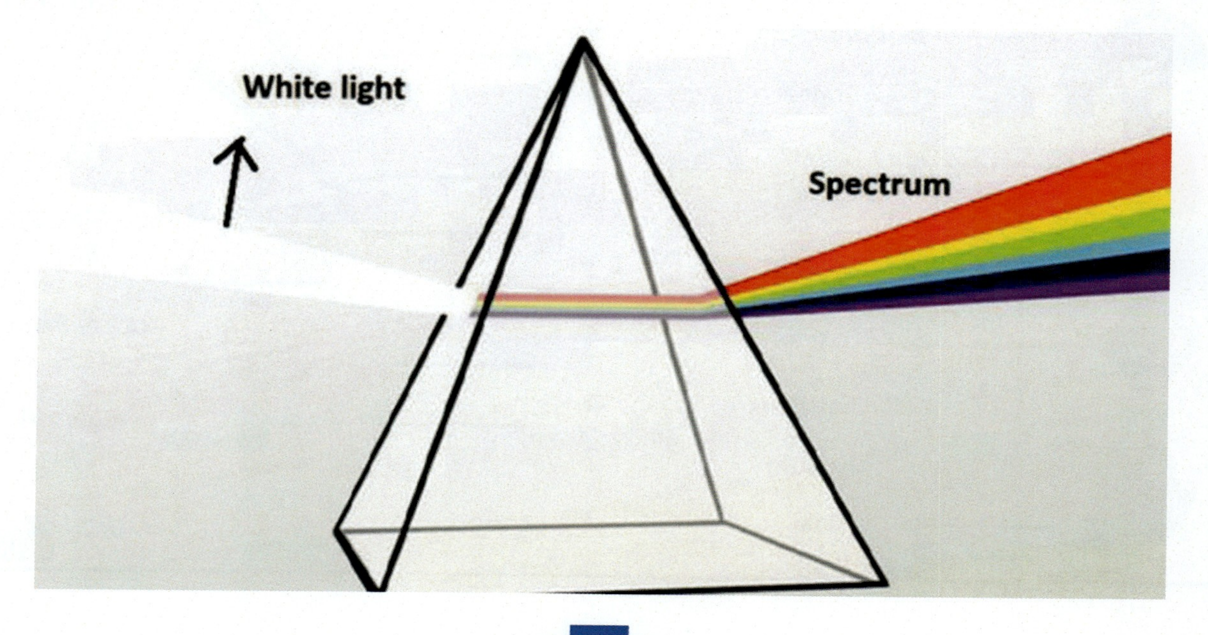

Luster is the classification of the way a mineral reflects light; a prism shows light in a rainbow form.

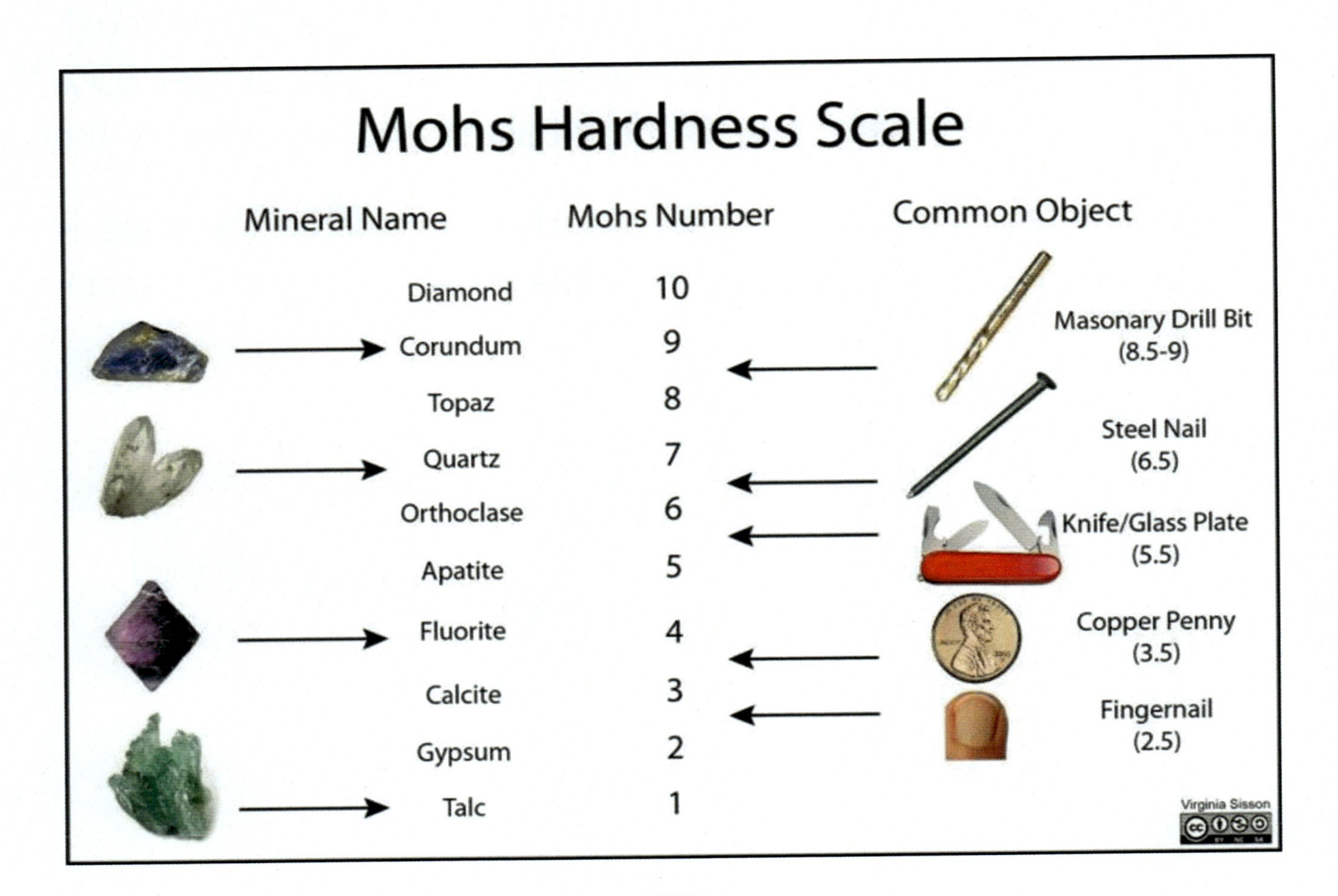

Granite

is a rock that was used in the past for decor, such as the pharaoh sphinx, which was made of smoothed granite. Many statues were made from granite so many mines ran out. When this happened, the renaming granite was made into an owl with a miniature human at its feet, they called it the granite god, which they used to beg for more granite.

As you're probably thinking, it did not work, but that did not stop them from trying. They used granite for writing, they would carve into the granite to make sacred symbols to their fallen leaders or pharaohs, then in the making of the pyramids, they also carved words to tell their ancestors to rest in peace or things they did not tell them when they were alive.

They also used it for masks that they did not wear but put in pyramids. Each one resembled a fallen leader in different ways; some came with a sacred word; others came with money glued with clay to resemble greed and wealth.

Basalt

is the most common rock to have ever been found in the Earth's crust, it makes up 70% of the Earth's crust! The Aztec calendar was also made from basalt.

Basalt can be found in dead volcanoes because during their destruction, columns of basalt line the bottom of the volcano.

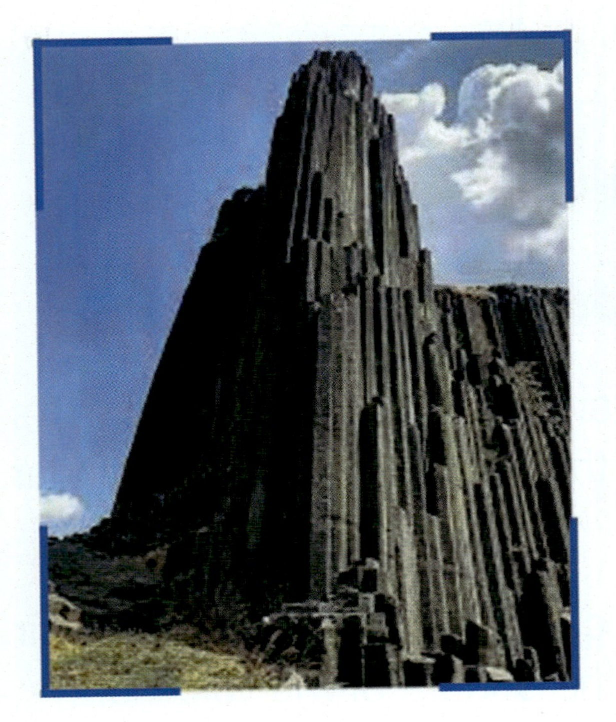

This could also be replaced by molten rock when cooled, but basalt is usually found near lava, so it is a result of a forming metamorphic rock.

Pegmatite

was used to make dye to use to write before the modern-day marker or pen. Speaking of pens, the tip of a modern pen is often made from pegmatite, but the ink is just regular dye. In the past, scammers used smooth pegmatite to make people think it was silver.

Then later, pegmatite was classified as rare, but only for a short time as people realized it was useless as currency.

Tuff

was first found in the ruins of Pompeii. People found out that there's a chance that when rocks cool, some might turn to Tuff. Other examples of tuff are metamorphic rocks.

In the ruins of **Pompeii**, people found a dog statue made of aged tuff, which meant tuff

was used before the volcano erupted and was not only made from the volcano in Pompeii. Tuff is still used in buildings today.

Pumice

is often turned into an oval like shape to be scrubbed on the foot or other places to scrape off dead skin.

It was also used in architecture. Many people used metal to make domes, but metal is very heavy, which is good for protection, but it was bad at making larger domes.

So, they used pumice to make a strong, lightweight dome that did better than the metal, and for some reason, was cheaper.

Obsidian

is a black rock that can be found in mountain sides, and on rare occurrences, in deep valleys, but not ones with clay or red sand.

Obsidian was used for masks for women and sometimes art in the form of statues.

It can be found in small clusters called tears. You should be careful when holding obsidian because it can have sharp corners that can cut skin.

Obsidian was used fo and necklaces because of its look and color, so bold, it's almost impossible see any other color in obsidian but its blackness and the reflection of the light.

Dolerite

is a rock that was used to build Stonehenge, one of the wonders of the world.

Sometimes gold can be found in dolerite, which was something many people did not believe until they uncovered a 2,000-foot golden dolerite seam in Australia.

Limestone

is a rough rock that was used in the making of Giza, the largest pyramid in Egypt, which is one of the 7 wonders of the world. It was also used for all the other pyramids beyond its international border! If that did not impress you, what if I told you the Empire State Building was made of an estimated 20,000 tons of limestone! You're still not impressed?

Then this will, the Tower of London was made of limestone, and the tower is big, they shipped all this limestone with a boat. And you're still sitting there yawning, then this will wake you up the Leaning Tower of Pisa in Italy was made of limestone.

And if you're the most uninterested person out there, then let me tell you, the White House, the home of the president is made of limestone!

Breccia

is mostly left over material while mining because it doesn't really serve any use, but that doesn't mean it does not have a cool backstory.

Breccia is found in places of geological activity, or where it was. For example, many sinkholes may be filled with stones and breccia.

Speaking of stones, pebbles and stones can be found cemented together in breccia because it is a sedimentary rock and is really clusters of itself stirred up in sinkholes or other ruins, which means one hard strike of a hammer will reveal all the little clusters of Breccia.

Shale

is the lowest rock on the Mohs rock hardness list since shale is very easy to break. People sometimes use it for pottery instead of clay, which takes a long time to harden, while shale can be cemented together, and they can use the clay to cover the little cracks.

Shale is a sedimentary rock which is shown by all the layers and little cracks in its formation, and on the rock, we can see that little clusters were cemented onto it during its formation.

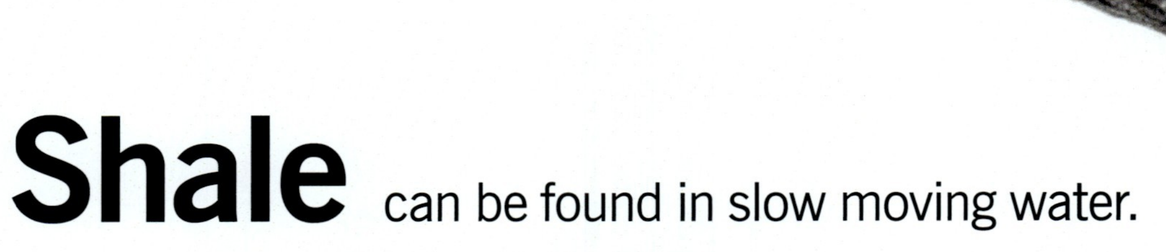

Shale can be found in slow moving water.

Fossils

Fossils are the remains of animals that lived in the past. They decompose into the ground, and all that's left is the bones of the animal, over time, this turns into a fossil.

Sometimes plants and bugs can imprint themselves in the fossil which allows us to see the habitat the animals lived, ate, and slept in.

Scientist can see how the plant or animal died by looking at the bones. Most of the time, if there are fractures in the bone, it would probably mean that a natural disaster killed it, or an animal ate it.

PLANTS

Some plants can only be seen under a microscope and others are super tall, but all of them have cell walls, and almost all of them have chlorophyll to help produce their own food from the heat and water in the ecosystem the plant lives in.

Plants are very versatile and can adapt to extreme conditions.

For example, because mangrove **swamps** are warm and have limited fresh water, plants must evolve to use salt water instead of fresh water, which most plants live in. They also have pigments that help with photosynthesis.

In the past, the only plants were all microscopic and algae, and they could not live on land. Thousands of years later, some ferns and other plants started **Sprouting**, and they were able to develop many structure habits in the growth, so they don't grow crooked.

Different

plants need different soils, some are humid, others are dry, or very salty, but all plants can usually live in only one kind of soil.

Many farmers are very happy to see worms in their farms because it shows that the soil is humid and has a lot of minerals the worms need to eat.

Worms are very sensitive to the sun, so they like damp wet places, like the ground that gets good amounts of rain every week. If the worm is in the desert, it digs deep into the ground to where water flows to reach damp places.

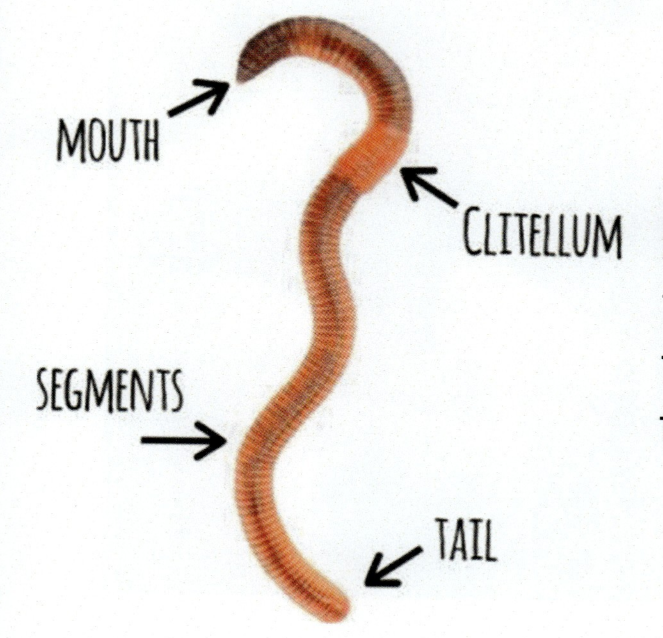

A worm could not live in the continent of Antarctica, even though it's wet, there's no place for it to call home.

Have you ever planted a tree?

If you have, have you seen its progress?

It's amazing to see a little seed turn into such a beautiful tree. When a seed or nut falls to the ground, it sinks into the **soil** and starts to sprout, then it pokes out of the ground with two little leaves on the stem. If it's a type of fruit tree, the plant starts growing flowers when the bark and wood comes.

After this, it will start growing different fruits according to the seed. The tree makes seeds that are in the fruit, then when a person eats them and puts the seed in the soil, it will make another fruit tree!

The **tree** made an offspring, a copy of itself that will soon sprout and make more offspring and evolve to work in harsher weather.

Cells in our bodies do the same thing, when you grow older your skin will get longer and looser.

This happens when cells multiply, but the cells get smaller because the cells divide themselves to increase the area.

Many plants have different ways to get water, some get water by transportation through the cells, others use small veins to transport water. Scientists use the definition vascular and non-vascular when comparing these differences.

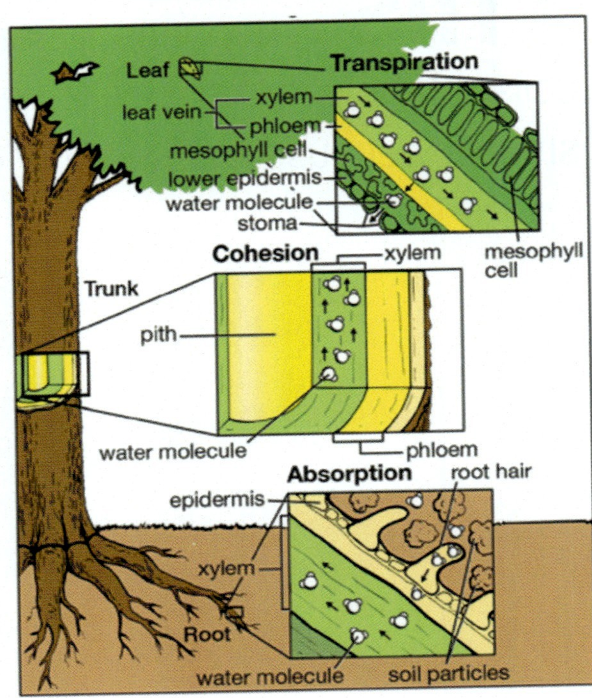

You can use a Venn-diagram to organize these simple studies to compare large amounts of data into a reliable source to check when needed.

Vascular

means the plants use veins to bring water, inside these veins are 3 tissues that help with this, they are Phloem, Xylem, and Cambium, but there is a difference, some of them carry water while the

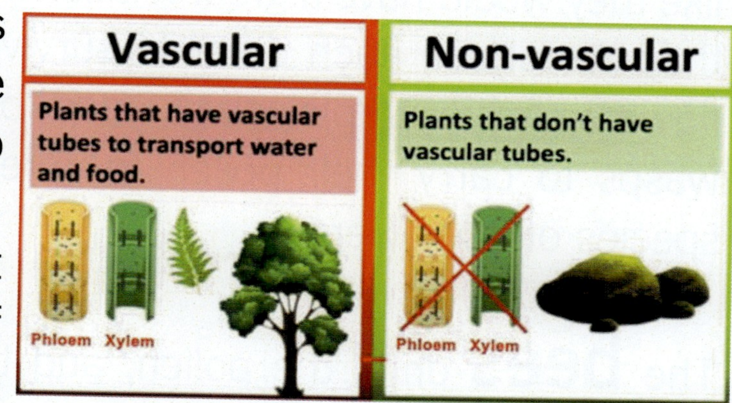

other veins take in the sustenance needed for the food produced by the plant. Flowers are plants, but do not grow food, they live from the Earth's wind and animals.

They produce seeds like fruit trees. You could use a diagram to study these similarities and differences.

Plants and animals don't look like they would have a great relationship, but they do rely on each other to survive. Like flowers, for example, they need bees or wasps to carry their pollen to spread the species of its flowers.

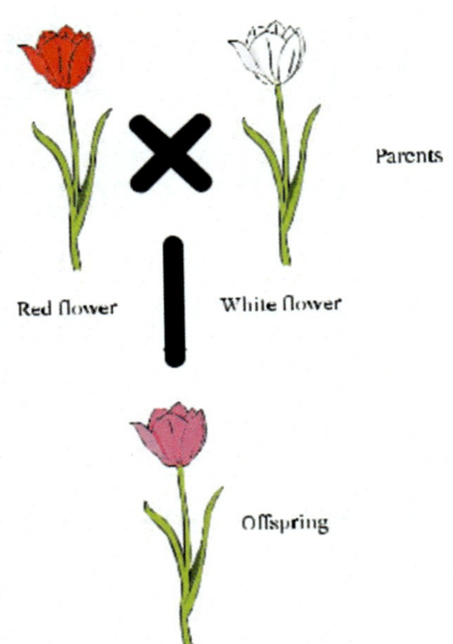

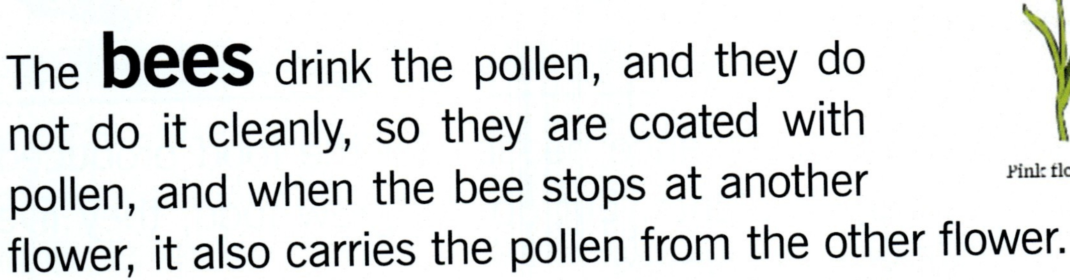

The **bees** drink the pollen, and they do not do it cleanly, so they are coated with pollen, and when the bee stops at another flower, it also carries the pollen from the other flower.

This is how flowers bond and make offspring, the doner that the bee took pollen from is the male in this situation, and the female is the flower that gets the pollen.

The flower is then able to make seeds now from the pollen, so when it dies, the seeds will fly away to sprout, and it happens all over again.

HAVE YOUR PARENTS EVER TRIED TO GET RID OF A

BEEHIVE?

What probably happened was the bees swarmed to protect the hive. They do their best to protect their queen bee, the mother of the bees.

Some bees protect, some build, others scavenge and make honey for the mother of the bees to keep reproducing.

If the hive is damaged, it would have killed a lot of the bees and some builder bees so they cannot rebuild.

Plus, they don't play easy, one sting and you're crying to your mama. Male bees cannot sting, only the females can, but that does not mean you should learn the difference and start messing around with them.

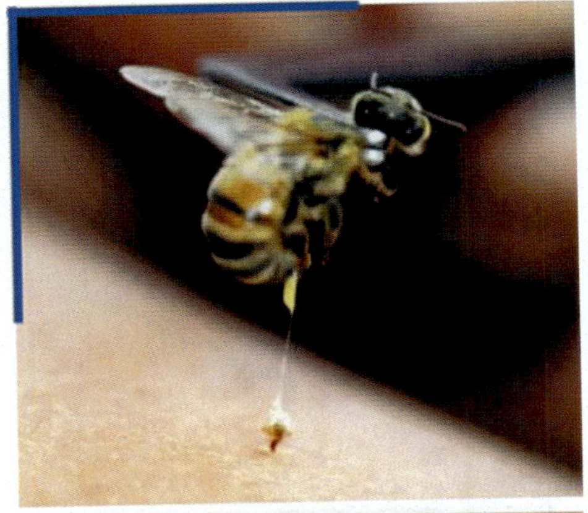

Bees can alert others of a threat, so if you want to do that, lie low for a couple of minutes, oh, and do not go outside.

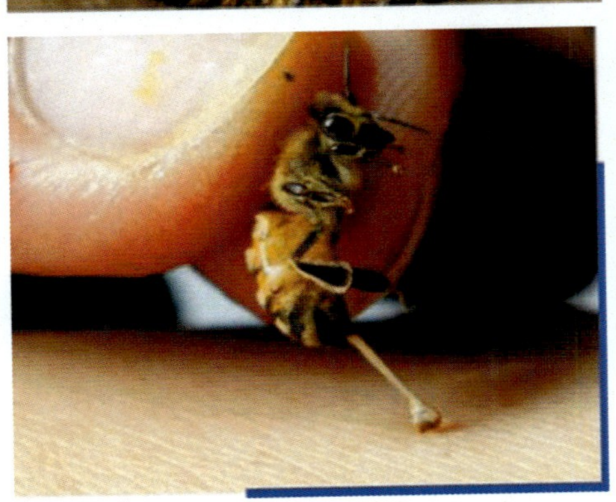

Many acres of trees have been cut down for buildings and space to accommodate the expanding population.

This can release a lot of carbon **dioxide** into the atmosphere and can fill the sky with carbon instead of air.

Many critters that live in the trees, including bees, squirrels, birds, and tiny slugs that hide in the bark, must find another tree, and then that one gets cut down, and it becomes a cycle.

Many It can get worse when many workers clearout Forests, like a species of monkey **almost** died off because a tree, their home, was cut down. Now, **scientists** are quarantining it until it reproduces.

The Amazon

rainforest, the biggest forest in the world, is also being cut down, and so you know, they don't **replant** the seeds, and most of the wood is not even used, so they burn it to turn into fossil fuels that pollute the air.

LOVE OUR PLANET..

Printed in the United States
by Baker & Taylor Publisher Services